668

A MATERIAL WORLD

It's PLASTIC

KAY DAVIES and WENDY OLDFIELD

Wayland

A MATERIAL WORLD

It's Glass It's Plastic
It's Metal It's Wood

Editor: Joanna Housley
Designer: Loraine Hayes

First published in 1992 by
Wayland (Publishers) Ltd
61 Western Road, Hove
East Sussex BN3 1JD, England

British Library Cataloguing in Publication Data
Davies, Kay
It's plastic. – (A material world)
I. Title II. Oldfield, Wendy III. Series
668.4

ISBN 0 7502 0363 3

Typeset by Kalligraphic Design Ltd,
Horley, Surrey
Printed and bound in Belgium by Casterman S.A.

Words that appear in **bold** in the text are explained in the glossary on page 22.

IT'S PLASTIC

Plastic is a modern material that is made from oil and natural gas. If you look around you, you will find many things that are made from different types of plastic. It can be made into hundreds of different shapes, including flat sheets and long, thin threads. It can be soft, like fur, or hard, like food containers. Plastic is useful because it can keep things waterproof and airtight. We also use it for decoration, because it can be made in all sorts of colours and shapes. This book will show you some of the many different uses of plastic in the world today.

Plastic can
be made
into any
shape we
like.
Its bright
colours
make it
fun to use
and wear.

4

The hard handle and soft bristles of this brush
are both made from different types of plastic.

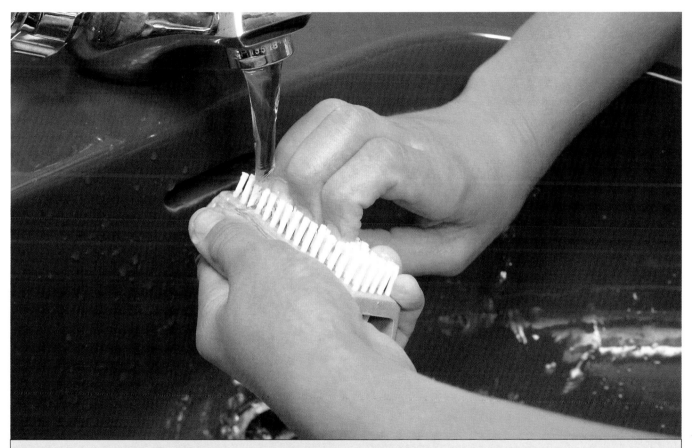

What other plastic objects are there in your
bathroom?

Thin sheets of clear plastic are useful for wrapping up food to keep it fresh.

We can use plastic bags to carry our heavy shopping in. Sometimes we can even see inside.

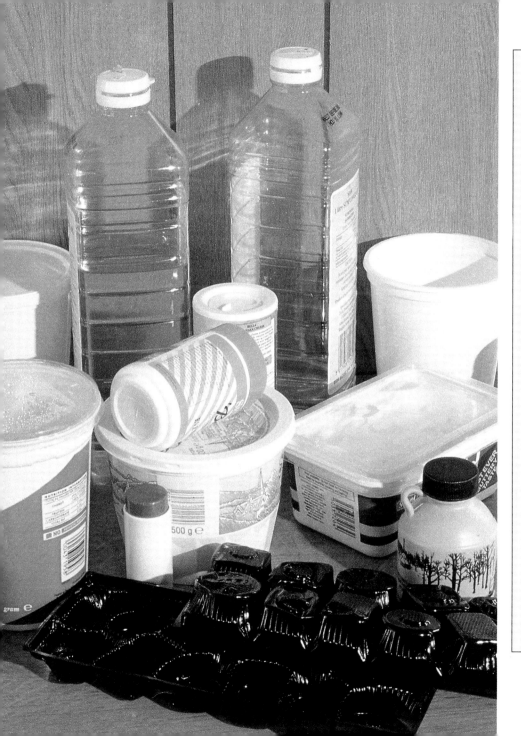

Plastic bottles,
boxes and
tubs protect
our food
and drink.
They help
to keep
things fresh.

They are
light and
strong,
and won't
break if we
drop them.

These toys feel soft when you stroke them.

Their fur is made from lots of very thin plastic threads.

Masses of small plastic balls fill the **ball pond**.

They move like water around us and make it easy
to exercise our bodies.

On a hot, sunny day it is fun to play in the water.
The plastic paddling pool and toys blow up like balloons.
They won't let water through their sides.

The huge **inflatable** is full of air. It's a bouncy playground where we can throw ourselves about without getting hurt.

Plastic can be made into thin threads called **synthetic fibres**. We can **weave** them into cloth to make comfortable clothes.

Are you wearing any plastic clothes?

Plastic rain-wear keeps us warm and dry. Rain bounces off and cannot wet our clothes and skin.

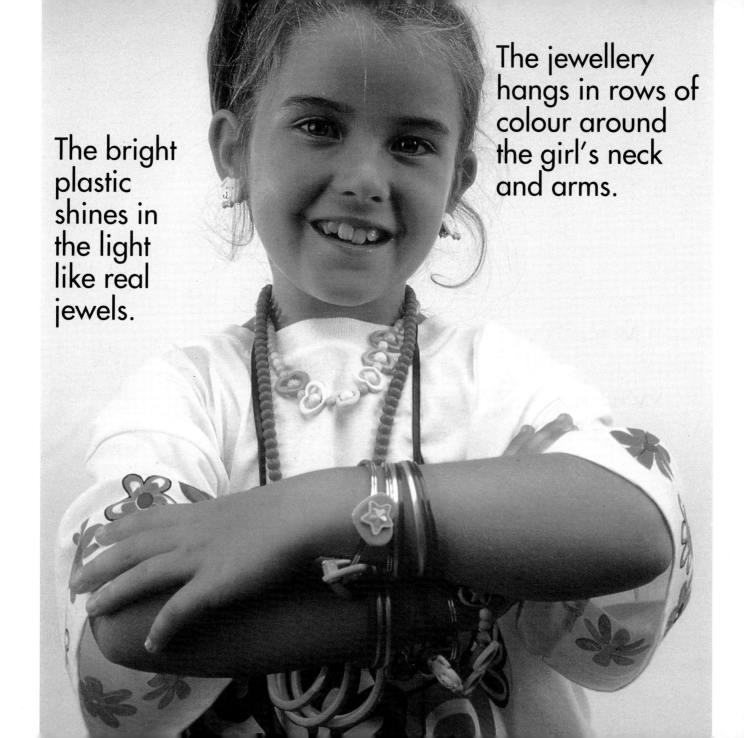

The bright
plastic
shines in
the light
like real
jewels.

The jewellery
hangs in rows of
colour around
the girl's neck
and arms.

The plastic
fruit looks
real.
At the
party
it makes a
light, fun
head-dress.

The helmet protects the firefighter's head from knocks and bangs.

The plastic **visor** is as clear as glass. He can see through it, and it will protect him from the heat of a fire.

We do not need to use the skin of dead animals to make things to wear. We can make handbags, shoes and belts from soft plastics that look and feel like **leather**. We can even wear pretend fur coats.

We like to talk to our friends on the telephone.
The plastic casing keeps us safe from the dangerous electrical wires inside.

It's easy to type into the computer on the plastic keys. You can see the game on the computer's screen.

We take plastic cups and **cutlery** on a picnic because they are light and don't break easily.

But they are a waste of plastic if we only use them once.

Old plastic can be collected. It will be cleaned and made into new bottles and bags.

GLOSSARY

Ball pond An area like a swimming pool which is filled with small plastic balls.

Cutlery Tools used for eating, such as knives and forks.

Fibres Thin threads of material, like hairs.

Inflatable A specially designed plastic shape that can be filled with air.

Leather The dried skin of an animal which is used to make shoes, clothing, belts and bags.

Synthetic Not natural.

Visor A see-through shield to protect the face.

Weave To make cloth by crossing threads over and under each other.

BOOKS TO READ

Plastics by Terry Cash (A & C Black, 1989)

Resources Control by Alexander Peckham (Franklin Watts, 1990)

Recycling Plastic by Joy Palmer (Franklin Watts, 1990)

Plastic Raincoat by Wayne Jackman (Wayland, 1989)

Some books in the following series may also be useful:
Simple Science (A & C Black)
Starting Science (Wayland)

TOPIC WEB

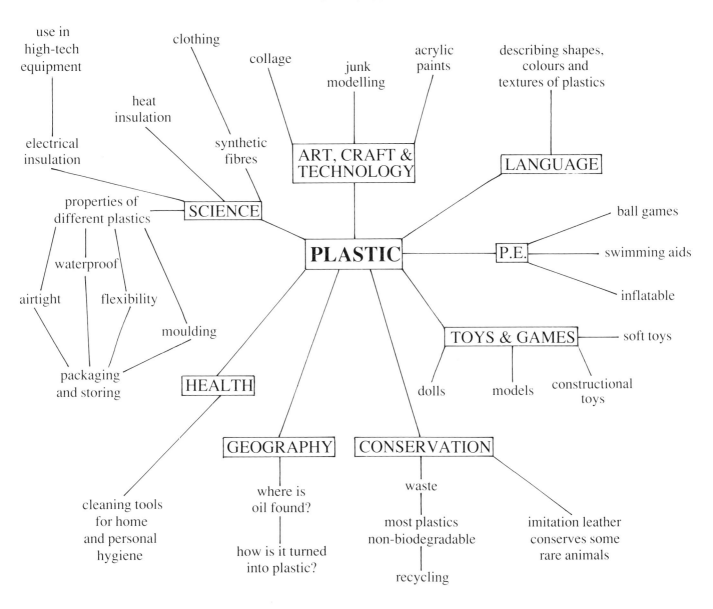

use in high-tech equipment

clothing

collage

junk modelling

acrylic paints

describing shapes, colours and textures of plastics

heat insulation

electrical insulation

synthetic fibres

ART, CRAFT & TECHNOLOGY

LANGUAGE

properties of different plastics

SCIENCE

PLASTIC

ball games

swimming aids

P.E.

inflatable

waterproof

airtight

flexibility

moulding

TOYS & GAMES

soft toys

packaging and storing

HEALTH

dolls

models

constructional toys

cleaning tools for home and personal hygiene

GEOGRAPHY

CONSERVATION

where is oil found?

waste

how is it turned into plastic?

most plastics non-biodegradable

imitation leather conserves some rare animals

recycling

23

INDEX

Picture acknowledgements
Chapel Studios 4, 6 (both), 7, 12 (inset), 14, 15, 17, 20; Environmental Picture Library 21 (Vanessa Miles); Eye Ubiquitous cover (top and right, Paul Seheult); Hutchison Library 9; Tony Stone Worldwide 10 (Peter Cade), 13 (Nicole Kitano), 16, 18 (David Sutherland); Wayland Picture Library 5 (Tim Woodcock), 11 (Chris Fairclough), 12 (main pic, Tim Woodcock); Tim Woodcock 19; ZEFA 8.